COLD FUSION

One of
Science's
Impossible
Theories

COLD FUSION

One of
Science's
Impossible
Theories

Dr. Austin Mardon | Poojitha Pai | Maryleen Geronimo | Viveka Pimenta
Wenteng Hou | Amanda Rande | Alexander Martin | Krish Vig

GM PRESS

Typeset and Cover Design by Brett Boyd

ISBN: 978-1-77369-501-3
Golden Meteorite Press
103 11919 82 St NW
Edmonton, AB T5B 2W3
www.goldenmeteoritepress.com

CHAPTER 1

The History of Cold Fusion

Poojitha Pai

Over thirty years ago, in a press conference at the University of Utah, two electrochemists proclaimed that they had attained a 'sustained nuclear fusion reaction' at room temperatures. The media quickly jumped on this unprecedented development in energy research, calling this process 'cold fusion' – the production of energy the same way the stars did, but at room temperature (Greshko, 2019). If true, this discovery could change the landscape of energy production overnight, affecting millions around the world. However, unlike the discoveries of quantum physics or the structure of DNA, the story of cold fusion is not rife with Nobel prizes and scientific achievement. Instead, it is rooted in controversy and disappointment that plagues this topic to this day.

What is Nuclear Fusion?

Before we delve into the history of cold fusion, it is important to briefly understand the science behind it so that the gravity of this "discovery" is fully comprehended.

Atoms, in most simple terms, are formed by three main subatomic particles – electrons, protons, and neutrons. Protons and neutrons reside in the center of the atom, in a dense area of space known as the nucleus while the electrons orbit around them (Bohr, 1923). Nuclear fusion, in general, occurs where pairs of nuclei fuse together to form a new nucleus and different subatomic particles. The difference in the masses of the products and reactants causes the release of a large amount of energy that is described by Einstein's iconic equation $E = mc2$, where E is the energy, m is mass, and c is the speed of light. Currently, we depend on various resources such as fossil fuels for energy, which are non-renewable. Harnessing the power of atoms that are abundant on Earth would greatly help answer the growing demand for energy on the planet.

However, there are many obstacles in the way of achieving nuclear fusion, the most notable of which is the "Coulomb barrier". Since atomic nuclei are positively charged, they greatly repel each other when brought close together – similar to how the same poles of magnets repel each other – forming the basis of the Coulomb barrier. The only way to stop this repulsion is to make the nuclei hit each other at extremely high speeds, which only occur at high pressures and temperatures, as it is on the Sun (Greshko, 2019). As such, it is exceedingly difficult to achieve these conditions in controlled conditions in labs due the expensive resources required to achieve this. While we have seen these conditions being successfully achieved in nuclear weapons, they are extremely destructive and have not allowed for humans to harness or control this energy in a sustainable manner. Regardless, having to detonate nuclear weapons to light up a street lamp is not exactly ideal.

This is where cold fusion comes in. Unlike the "hot" fusion that occurs naturally in stars in extreme conditions, cold fusion is a hypothesized type of nuclear reaction that could possibly occur near room temperature using electrolysis. This was the kind of breakthrough that scientists dreamed of making – simple experimental set up, easy to measure results, and yet the power to turn the scientific world on its head. Two electrochemists, B. Stanley Pons and Martin Fleischmann, set out to make those dreams a reality.

Early Research

While Pons and Fleischmann were the first ones to fully "observe" cold fusion, they were not the first to attempt it. Thomas Graham, a British chemist primarily known for his pioneering work in dialysis, discovered that palladium, a heavy metal, could absorb hydrogen into its crystalline lattice in 1866. This would form a crucial part of Pons and Fleischmann set up nearly a century later (Graham, 1869). Fritz Paneth and Kurt Peters of the University of Berlin also experimented with hydrogen in palladium in 1926. It was defined by Nature as "a transmutation of hydrogen gas into helium, based on the use of finely divided palladium". This was one of the first reports of one nucleus (Hydrogen) turning into another nucleus (Helium). However, within a year, Paneth and Peters retracted their claim, stating that the observations may have been produced by the contamination of helium in the background (Krivit, 2009). This gave the world the first inkling of the controversy that would rack cold fusion in the next few decades.

In 1927, a Swedish scientist named John Tandberg also reported that he had fused hydrogen into helium in an electrolytic cell with palladium electrodes. However, due to Paneth and Peter's retraction as well as his inability to explain the physical process itself, his ideas were never too credited or publicized. After the discovery of deuterium, which is an isotope of hydrogen that contains two neutrons instead of one, he continued his experimentation. Unbeknownst to Pons and Fleischmann,

Tandberg's later experiments were very similar to the ones they attempted later on (Murdoch, 1989).

The Press Conference of 1989

Stanley Pons and Martin Fleischmann were an unusual pair. Pons was a quiet and modest man from a small town in North Carolina, USA, while Fleischmann was a boisterous European man who was old enough to be Pons' father. The two had met in the University of Southampton in England, where Pons completed his PhD and Fleischmann was a professor. Shortly after Pons had become a professor in the University of Utah, they both collaborated on multiple research projects (Cold Fusion: A Case Study for Scientific Behavior, 2012).

The idea behind their subsequent cold fusion experiments were based on the previous research that had come before this, including Fleischmann's own observations. In the late 1960s, he had observed that palladium could absorb exceedingly large amounts of hydrogen and deuterium, about 900 times its own volume. He then reasoned that the only way to fit this number of atoms into the palladium would be for the atoms to be "squished". This could be a possible means to fuse multiple atoms together to release energy (Cold Fusion: A Case Study for Scientific Behavior, 2012). He also knew that other similar experiments had reported the presence of helium after these experiments, which could imply a nuclear reaction taking place (Krivit, 2009).

To investigate this phenomenon further, he and Pons started running electrolysis experiments with palladium electrodes and deuterium in what they called a "fusion cell". This cell consisted of two pieces of metal, palladium and platinum, submerged in a container with heavy water (water in which hydrogen was replaced with deuterium). If they provided this cell with electricity, it would trigger electrolysis in which the heavy water would split to produce oxygen and deuterium gas, which would be absorbed into the palladium. Based on their previous understanding of palladium, they hypothesized that, eventually, so many deuterium molecules would be forced to be so close together that fusion would occur, releasing large amounts of energy that could be harnessed (Cold Fusion: A Case Study for Scientific Behavior, 2012). After weeks of seeing nothing, they found one morning that the experiment had gone ballistic the night prior. A substantial portion of the palladium cathode, which had a melting point of 1544°C, had fused and vaporized. Many parts of the lab equipment, including the lab bench, concrete floor, and fume cupboard were destroyed. Further details about the experiment are not very well known as Pons and Fleischmann had been working in secret. Regardless of the details, they were convinced that something had happened that allowed the experiment to reach extremely high temperatures or made it explode, which spurred them on to carry these experiments in secret for several more years until they eventually ran out of money (Krivit, 2009).

In 1988, Pons and Fleischmann applied to the United States Department of Energy to fund their experiments, finally clearing some of the mysteries that their experiments were shrouded in. Before this, most of the funding for their relatively small-scale experiments came out of their pockets but still cost over a hundred thousand dollars. For this application, they needed to provide a grant proposal to the Department of Energy for peer review. One of the reviewers chosen was Steven Jones of Birmingham University, who was working on muon-catalyzed fusion. Muon-catalyzed fusion was a known method of inducing nuclear fusion without the extreme temperatures and pressures required for "hot" fusion (Homlid, 2019). He reached out to them with the aim to collaborate on this topic but was refused. As such, he was competition for Pons and Fleischmann. The only difference was that Pons and Fleischmann were convinced that their experiment would be useful for commercial applications while Jones's research was less commercially inclined. Despite their shared belief that cold fusion was possible, Jones' results did not match Pons and Fleischmann's findings. Nonetheless, when they realised that they had remarkably similar research aims, both parties agreed to publish their results on the same day to Nature so as to not overshadow one another's achievements (Cold Fusion: A Case Study for Scientific Behavior, 2012).

Unfortunately, this agreement was broken just a few days later. Pons and Fleischmann held a press conference to publicize their results. In this press conference, little information of substance was given, but it garnered the interest of those within and out of the scientific community. Other ground-breaking discoveries such as high-temperature superconductivity in the same decade made the scientific community more open to unexpected new research like this. The announcement also could not have come at a more perfect time – the 1973 oil crisis was still fresh in many individuals' memories, the rise of global warming awareness, and an oil spill that occurred the day after the press conference. All of these factors together influenced the way that many held this discovery in high regard even with a lack of any real proof.

In their announcement, the two scientists made a few claims:

1. *They had measured massive amounts of energy in the form of heat.*

2. *There was a novel nuclear process at play which defied the normal signs of nuclear processes – no strong neutron emission or gamma radiation.*

3. *They had sustained fusion of deuterium in their experiment.*

All these claims were based on heat measurement data that they had recorded along with low levels of some nuclear products such as tritium and low levels of gamma radiation. They also released their paper to the Journal of Electroanalytical Chemistry prior to the conference, whose editor sped up the peer review process as "special treatment" due to the gravity of the findings, especially since they claimed that the energy release measured was many times larger than any known chemical reaction would be able to explain. Due to the sped up peer review, many flaws in the paper were missed (Krivit, 2009). In the rush to publish, the high standards and rigor of scientific publishing was compromised. Additionally, Jones, angered that their agreement was broken, submitted his paper to Nature after the conference (Lewenstein, 1992).

The Fallout

As is the case with any new discovery, especially one with consequences as vast as this one's, many scientists got to work on replicating the experiment. While they did not have an experimental protocol yet, many scientists in several countries attempted to replicate this study with extraordinarily little success. Instead of immediately discrediting cold fusion, many scientists thought that they may not have been replicating conditions for success perfectly. As a result, many of them asked to collaborate with Pons and Fleischmann, who refused everyone. With this continuous refusal, they were standing in their own way to proving the validity of their claims.

Moreover, critics at the Massachusetts Institute of Technology (MIT) found an error in the gamma ray data, who then reported these inconsistencies publicly. Many esteemed professors at MIT went as far as initially calling Pons and Fleischmann frauds before retracting their allegations and instead suggesting that it was probably a mistake. Douglas R. O. Morrison of the European Organization of Nuclear Research (CERN) took a much stronger stance, having confidence that cold fusion was perhaps a "delusion" of Pons and Fleischmann. Most of these speakers were gathered at the American Physical Society (APS) conference of 1989, making this the height of the assault on the two scientists' work as well as character.

After the APS conference, the backlash and controversy had become impossible to ignore. The Department of Energy put together a panel, which included Nobel Laureate physicists, to review cold fusion. Six months after the panel had been put together, it was reported that the cold fusion experiments had not, in fact, produced fusion products in the expected amount. After looking at multiple labs who had tried to replicate the results, and observing the extremely high rate of failure, the panel was not convinced that this technique would result in the production of any substantial amount of energy (Krivit, 2009). Over a year after the press conference was held, the scientific process had been able to sort through all of the evidence and conclude that cold fusion had not yet been attained. Nature also published a major article claiming no evidence of cold fusion in the original Pons-Fleischmann experiment, thoroughly putting an end to this dream (Lewenstein, 1992).

The Legacy

The press conference in which Pons and Fleischmann publicized their findings sparked one of the most heated scientific controversies in recent times and was filled with resentment, hostility, and bitterness. Errors that ultimately led to the downfall of this scientific endeavor could have been caught early on. From Pons and Fleischmann who published their results without the proper checks and balances and by breaking their agreement with Jones, to the editor who thought that their paper deserved special treatment, and to those scientists who were able to reproduce these results in their labs only to later retract them, the shortcuts taken in the rush to publish heavily compromised the high standard that the scientific community holds itself to. Thousands of hours and millions of taxpayers' dollars were squandered on this dream of cold fusion. Additionally, the effect this fiasco had on the general public is also worrying, as the sensationalism and the accusations of fraud may have damaged the trust they have in the scientific community, which is something that is long-lasting and plagues us even in the middle of pandemics. It is, however, a small reassurance that science, despite all the drama and flaws, still had worked i.e. most of the flaws were caught and the mistakes rectified as much as possible. Nonetheless, the powerful idea that we could harness the energy of fusion still spurs some scientists to investigate the possibility of cold fusion. But with this damaged legacy, they will need to find solid, replicable evidence to support their views to convince the rest of the world.

CHAPTER 2

The Varieties in Nuclear Fusion

Maryleen Geronimo

As our world continues to expand and evolve, so has our demand for energy. This is where nuclear fusion comes into play. Nuclear fusion can be simplified as the "process by which certain light atomic nuclei react with each other to release energy" - a reaction where its supplies are abundant (Morse, 2018). Due to this abundance of supplies, the leading area of research today is the hunt to successfully harness this energy, in an attempt to fulfill our current demands for energy. In other words, many scientists believe that as a result of advances in fusion technology, the world can be saved from a future environmental calamity. (Nuclear fusion promises a clean, green energy future, 2020).

Nuclear fusion has many advantages, including its reliability, efficient land-use, being environment-friendly, and its ability to be implemented anywhere around the world (Benefits of Fusion, 2021). This technology is reliable as it is independent of weather. In other words, regardless of the conditions, nuclear fusion will continue to provide energy on-demand. Second, nuclear fusion results in zero greenhouse gas emissions, with its only production being helium, unlike other sources of energy such as fossil fuels. Finally, nuclear fusion requires fuel that can be easily sourced from seawater, a resource that is extremely abundant. This feature allows the implementation of fusion energy anywhere in the world, providing countries with energy independence.

Nuclear fusion has always been a dominating reaction in our universe, particularly as it is the prime reaction that powers the Sun and stars (EUROfusion, 2017). Even so, the science behind nuclear fusion was not proposed until the 1920s, by the British astrophysicist Arthur Eddington. Eddington theorized that the stars are able to extract their energy from the fusion of hydrogen, resulting in the production of helium. This theory, originally published in the 1926 edition of the Internal Constitution of the Stars, can be credited to laying the groundwork for current theoretical astrophysicists (EUROfusion, 2017). Following Edison's theory, many others took interest in the subject, including Robert Atkinson and Fritz Houtermans. These two scientists were able to present the first calculations that demonstrated the rate at which nuclear fusion occurred in the stars. It was not until the 1950s when researchers began to explore the possibility of reproducing the reaction of nuclear fusion on Earth. In 1950, a design for a "magnetic confinement fusion" device called the tokamak was proposed by the scientists Andrei Sakharov and Igor Tamm (EUROfusion, 2017). In the following year, Lyman Spitzer designed a similar device called the stellarator. Upon further experimental research, it was determined that the concept of the tokamak was more efficient. As research surrounding nuclear fusion continues, the difficulties in relation to harvesting fusion energy have become apparent for many scientists. Therefore, in 1973, a number of European countries agreed to work together, creating the Joint European Torus (JET). The JET is a UK-based research facility, labelled at the time as being the "largest and most successful fusion experiment in the world" (EUROfusion, 2020). During the Geneva Superpower Summit of 1985, the creation of ITER (International Thermonuclear Experimental Reactor), a collaborative international project aiming to peacefully develop fusion energy, took place. Fast forwarding to the 1990s, JET was able to run the first reactor in the world. This experiment used a "50-50 mix of tritium and deuterium" as fuel for the reaction (EUROfusion, 2017). In 1997, through the same process, JET was able to successfully set the current world record for a "fusion output at 16MW from an input of 24 MW of heating", resulting in a Q value of 0.67 (EUROfusion, 2017). It should be noted that a Q value greater than 1 indicates the successful achievement of fusion energy.

Nuclear fusions are categorized into two types of reactions; (a) reactions that maintain the number of protons and neutrons (b) reactions that involve the conversion between protons and neutrons (Nuclear Fusion, 2021). The first type of nuclear fusion is observed in "practical fusion energy production", whereas the second type is often observed during the "initiation of star burning" (Nuclear Fusion, 2021). Amongst these two categories, nuclear fusion is further broken down into varying types of fusions. In this chapter, we will discuss magnetic confinement, inertial confinement, magnetized target fusion, and hybrid fusion.

The Background Behind Nuclear Fusion Technology

As mentioned, nuclear fusion naturally occurs in the Sun. It can do so because of the Sun's ability to provide massive gravitational forces, creating an ideal environment for fusion to occur. Unfortunately, the same cannot be said about the Earth's conditions. In order to successfully replicate the same reaction on Earth, several difficult requirements must be met (Basic Fusion Physics, 2016). First, the fusion fuel (varying isotopes of hydrogen), must be heated to temperatures high enough to surpass the Coulomb barrier. Second, the reaction must remain in an extremely pressurized environment, providing a high enough density to produce an acceptable fusion reaction rate. The last requirement for a fusion reaction is a suitable distance between the ions, including a long enough confinement to avoid possible cooling. In conclusion, for an artificial fusion reaction to take place on Earth, the ions (fusion fuel) must be heated and placed into a high-density environment, where they are close enough in proximity to fuse together. The majority of difficulties surrounding the experimentation of controlled fusion research programs is the "ignition phase" (Nuclear Fusion: WNA - World Nuclear Association, 2021). In order for ignition to be achieved, a sufficient number of fusion reactions must occur to become a "self-sustaining" reaction, thus only needing the addition of fuel to continue the process. In surpassing the ignition phase, the reaction is able to provide a net energy yield.

According to the most recent research, the ions that would most practically fuse together are the two isotopes of hydrogen – deuterium (D) and tritium (T) (Nuclear Fusion: WNA - World Nuclear Association, 2021). Deuterium is naturally found in seawater, therefore is an abundant energy source. On the other hand, tritium is naturally found in the upper atmosphere, primarily where "cosmic rays strike nitrogen molecules in the air" (US EPA, 2015). Due to the radioactive nature of tritium, usable quantities can be developed from lithium, using a fusion system (Nuclear Fusion: WNA - World Nuclear Association, 2021). Fortunately, lithium is an abundant element, mostly found in the Earth's crust, and less so, in the sea.

Amongst the experimental approaches to nuclear fusion, two methods stand out: magnetic confinement and inertial confinement.

Magnetic Confinement

Magnetic confinement fusion (MCF) is a method that utilizes strong magnetic fields to control and restrain fusion fuel, which in this case, is in the form of hot plasma (Nuclear Fusion: WNA - World Nuclear Association, 2021). Magnetic fields are the perfect tool for the confinement of D-T plasma (deuterium-tritium plasma). Due to the electric charges of the ions and electrons in the plasma, this tool enables the flow and control through the utilization of magnetic field lines. The objective of this tool is to prevent the movement of particles towards the "reactor walls"; preventing this contact will prevent the overall dissipation of heat, thus maintaining the rate of reaction.

In magnetic confinement experiments, the most widely used magnetic confinement shape is the toroidal shape, which is observed to be similar to a doughnut. An example of a confinement system that utilizes this effective concept is the tokamak, as we previously discussed. Following the original design of this system in the 1950s, several tokamaks have been constructed; the Joint European Torus (JET), the Mega Amp Spherical Tokamak (MAST), and the tokamak fusion test reactor (TFTR) (Nuclear Fusion: WNA - World Nuclear Association, 2021). As a result of its shape, the tokamak is designed to operate in relation to two sorts of fields, the toroidal and poloidal field. The poloidal field is produced by using a variety of horizontal coils, located outside of the toroidal structure. On the other hand, the toroidal field is produced using coils that are evenly located throughout the reactor.

Although the tokamak is an effective confinement concept, other alternative concepts are being developed, such as the stellarator and the reversed-field pinch (RFP) ("Fusion reactor - Mirror confinement | Britannica," 2021). The stellarator was first designed by Spitzer in 1951 but labelled inefficient in comparison to the tokamak. In the 1950s, this was the case, as technology lacked the proper computer modelling techniques, preventing the calculation of its accurate geometry (Nuclear Fusion: WNA - World Nuclear Association, 2021). The advantages to this concept is its use of only external foils to create an effective magnetic field, while also providing greater stability for hot plasma compared to the tokamak. The only downside to this concept is the increased difficulties in relation to designing and building its complex shape. On the other hand, the RFP is overall similar to the tokamak, but utilizes weak toroidal magnetic fields instead.

Inertial Confinement

In more recent years, researchers came upon inertial confinement fusion (ICF). Inertial confinement fusion uses strong lasers on a "small pellet of fusion fuel" in order to heat and compress the pellet into extremely high densities (Nuclear Fusion: WNA - World Nuclear Association, 2021). In this concept, the laser (or ion beam) works to target and heat the outer layer of a pellet containing D-T fuel. This action would then result in an outward explosion of the contained fuel, while simultaneously "generating an inward-moving compression front" that both heats and compresses the inner layers of the fuel (Nuclear Fusion: WNA - World Nuclear Association, 2021). This process will then create a cascading chain of reaction that heats the surrounding fuel, this phase is known as the ignition phase. In addition, the amount of time required for this process to succeed is based on the "limited inertia of the fuel" (Nuclear Fusion: WNA - World Nuclear Association, 2021). On average, this process takes less than a microsecond.

The US National Ignition Facility (NIF) is an example of a large-scale inertial confinement fusion research device. This device is equipped with over 190 lasers, which can provide "more than 2 MJ of ultraviolet energy and 500 TW of peak power" (Nuclear Fusion: WNA - World Nuclear Association, 2021).

In a similar concept to inertial confinement fusion, the "Z-pinch" (also known as the zeta pinch) utilizes "strong electrical currents" to generate x-rays, which are then directed towards the plasma. This enables the compression and the heating of the fuel plasma.

Magnetized Target Fusion

Magnetized target fusion (MTF), also known as magneto-inertial fusion (MIT), is a concept of fusion that combines inertial confinement fusion and magnetic confinement fusion (Nuclear Fusion: WNA - World Nuclear Association, 2021). In the inertial confinement portion of this approach, the process of "compressional heating" is implemented. On the other hand, the magnetic confinement portion of MTF utilizes the processes that prevent the reduction of heat in the system (via magnetic field lines), while ensuring "enhanced alpha heating" of the plasma fuel (Nuclear Fusion: WNA - World Nuclear Association, 2021). Current scientists have grown interested in MTF as it requires less funding and is a simpler approach to nuclear fusion compared to previous designs.

Magnetized target fusion was first proposed by the United States Naval Research Lab in the 1970s (Delbert, 2021). Currently, both public and private companies have taken interest in MTF. A notable Canadian private company that is known for its experiments regarding magnetized target fusion (MTF) is General Fusion, widely known as the company "backed by Amazon's Jeff Bezos" (McGrath, 2021). General Fusion has been known to be very highly anticipated, as they had been reported to have raised over $100 million in their latest round of funding. General Fusion's process of MTF is described as follows (McGrath, 2021). The fuel plasma is placed into a cylinder container, which is located in the middle of a liquid metal wall. Pneumatic pistons are then utilized to compress and heat the fuel plasma, resulting in a fusion reaction. This process results in the release of vast amounts of heat, which is transferred through the liquid metal, boiling the surrounding water. The steam that results from this process will then be used to power a turbine (McGrath, 2021). A key advantage to the system used by General Fusion, is that all of the system's components already exist amongst our current technology.

Hybrid Fusion

Hybrid fusion combines the concepts of nuclear fusion and nuclear fission. Nuclear fission, the opposite of nuclear fusion, can be defined as the process by which an "atom splits into two or more smaller atoms, most often the result of neutron bombardment" (Nuclear Fission | Boundless Chemistry, 2021). Before discussing the process of hybrid fusion, we must understand the role of a "blanket" that surrounds the core of the reactor. During a fusion reactor, neutrons will be produced by the fusion of deuterium (D) and tritium (T). These neutrons will then be absorbed by a blanket (containing lithium) that surrounds the core, resulting in the conversion of lithium into tritium. This product can then be further used in the fusion reaction.

In a hybrid fusion, the fusion portion of the reaction will provide neutrons to the blanket. The captured neutrons in the blanket will influence fission reactions to take place, resulting in more neutrons. In this case, fission reactions occur because of neutron bombardment, as neutrons continue to strike one another (Nuclear Bombardment Reactions, 2021). To conclude, a positive feedback loop will occur, as an increasing number of neutrons will induce more fission reactions to occur. The advantages to this type of fusion are that the development of new materials, that can withstand neutron bombardment, is not necessary. Whereas, in a typical fusion system, new materials are required for the production and maintenance of the "blanket". In addition, the fusion portion of the hybrid fusion system would not need to produce as many neutrons, compared to the typical fusion system. As mentioned, this is a result of the positive feedback loop that ensues because of the nuclear fission reactions occurring. Finally, the fusion reactor used in the hybrid system will not have to be as large, compared to a fusion-only reactor (Nuclear Fusion: WNA - World Nuclear Association, 2021).

Conclusion

In conclusion, nuclear fusion is an extremely important topic of discussion. The use of such advanced technology has the power to significantly reduce our environmental impact, which is especially critical as the demands for energy continue to increase. Besides this, nuclear fusion has several other advantages such as being independent from weather, efficient with land-use, environment-friendly, and can be readily used anywhere in the world. In addition, there is great importance to studying nuclear fusion today, as it is a highly anticipated energy-source that will most likely takeover the world in the following decades.

CHAPTER 3

What Cold Fusion Could Mean for Renewable Energy

Viveka Pimenta

Current State of Energy Usage, Clean Energy, and the Environment

With each passing year, the effects of climate change become more severe as further greenhouse gases (GHGs) are produced by our human activities, trapping heat in our atmosphere. Worldwide energy consumption has only increased, and projections for the planet are looking more and more dire each year. Rising temperatures prompt more frequent and severe droughts and tropical storms while glaciers, sea-ice, and permafrost melt, increasing sea levels and releasing more greenhouse gases into the air (What are some of the signs of climate change?, n.d.). As a result, the need for clean, carbon-free, and renewable energy that can sustain humanity and protect ecosystems from deteriorating has become an increasingly urgent concern in recent decades.

Industrialization over the last century has driven the economy for fossil fuels, and the increased energy consumption has provided a growing market for the mining of carbon based energies like oil, coal, and natural gas, all of which have contributed to the dire state of the planet's climate today. Carbon-based energy is a non-renewable resource that harms the planet through its mining, production, and usage. In fact, 78% of worldwide GHG emissions are a direct result of human energy consumption and production (Government of Canada, 2017). GHG emissions like carbon dioxide and methane, which are released during production, increase the planet's temperature and drastically change weather patterns across the globe, placing people, animals, and plants in unpredictable habitats that are not meant to sustain such changes. The air becomes polluted as these energies are burned, causing people and animals to breathe in toxic products which adversely affect their health. The cycling of water through the air also becomes polluted as the environment worsens, creating acid rain. Even beyond warming concerns, every aspect of the environment is harmed by these non-renewable resources, from oil spills killing marine life to land habitat destruction resulting from mining procedures that change the landscape and introduce foreign chemicals into food sources for wildlife. Worst of all, these non-renewable resources will eventually run out, leaving humanity on a polluted, overheated earth with catastrophically damaged ecosystems, poor biodiversity, and no further energy to sustain our lifestyles (Green Tech Talk, 2019).

Many countries have transitioned away from using fossil fuels, particularly due to the 2015 Paris Agreement, which accelerated the shift towards an economy in favour of low global carbon emissions. Thanks to this international agreement, countries such as Canada have developed frameworks to phase out coal consumption, increase the tax on carbon pollution (thereby disincentivizing non-renewable energy choices), drive technological advancements to improve the state of the environment, and target the individual effects of climate change. Canada, in particular, has reduced coal consumption by 26% since 2000 and has set a target of decreasing GHG emissions by 30% by 2030 (Government of Canada, 2017).

These goals are made a reality through the use of renewable energy sources that do not produce GHGs, and actually generate energy at the rate of consumption or faster, such as wind, solar, hydro, and nuclear energies, to name a few. In fact, 2018 data attributed a whopping 82% of Canada's electricity to renewable sources (Government of Canada, 2017). However, that same year, the country's total primary energy supply (including transportation, heating, and industrial processes) was still mostly made up of non-renewable sources, with 76% coming from fossil fuels. Globally, there is also still a long way to go as 81% of the total primary energy supply is made up of fossil fuels (Government of Canada, 2017).

Canada's primary source of renewable energy is hydro, generating 60% of the country's power by converting the mechanical energy produced from moving water turning a turbine into electricity (Government of Canada, 2016). Other types of renewable energy resources include biomass, in which energy is generated from living organisms and biological materials like wood pulp, wind turbines which harness the kinetic energy of moving wind to produce electricity, solar power converting sunlight's energy into electricity using solar photovoltaic panels, and liquid biofuels which are liquids like ethanol or renewable diesel that get mixed with regular gasoline to reduce emissions (Government of Canada, 2016). There is also geothermal energy, in which the geology of the planet itself provides the energy via molten rock heating up water at the earth's crust, allowing the steam to be converted to electricity (Union of Concerned Scientists, 2013). Another rising star in the zero-emissions game is nuclear energy, which in the common vernacular, typically refers to nuclear fission, or the splitting of atoms to harness the heat energy byproduct of the process (Office of Nuclear Energy, 2021). Nuclear fission is the second largest renewable resource in the world, keeping the air clean while producing more power per unit of land than any other renewable resource, with 1000 megawatts of electricity being generated for every square mile, and still reducing waste due to nuclear fuel's high energy density and potential to be reused (Office of Nuclear Energy, 2021).

In the race to save the planet, these renewable resources have been advanced to a functional capacity and widely commercialized. However, even with their benefits, each of these resources has imperfections that limit their true sustainability. Selection of the best renewable energy technology falls to the following technical, environmental, economical and social criteria (Demirtas, 2013). The technical criteria encompasses the capacity for energy production, how long the technology has been in use, how reliable the energy source is, and how safe it is. Environmentally, impact on the ecosystem as well as GHG emissions have to be taken into consideration for the implementation of the resource. Economically, the cost of investment into the technology, operation and maintenance costs juxtaposed with service life, and the payback period all must be considered. Beyond these criteria, the resource must be socially acceptable and beneficial to the majority (Demirtas, 2013).

Currently, wind power is the renewable resource that meets the most criteria, as wind is unable to be depleted in quantity, is widely accessible to a large number of regions that experience high wind speeds, is lower in cost due to the simple technology and rapid development, and produces no pollution (Demirtas, 2013). However, it too has environmental impacts, such as the large amount of land required to host a wind farm. This creates challenges for the environment by disrupting habitats and harming the wildlife (Union of Concerned Scientists, 2013). Solar power harms the environment through habitat loss as well, and in the production of the technology, hazardous materials are released. Geothermal energy involves drilling, so it also impacts the habitats before any energy is even harnessed. Biomass power plants, while renewably produced, still utilize combustion to produce electricity, and thus release emissions and use excess water just like fossil fuel plants. Even hydroelectric and hydrokinetic power involves a range of environmental impacts (Union of Concerned Scientists, 2013). Nuclear fission power plants, while an incredibly clean, regulated source of energy, are still expensive to produce and generate long-lived radioactive waste that must be disposed of to prevent exposure, and this is when they are working perfectly. This is to say nothing of accidents at nuclear fission power plants, which create massive nuclear meltdowns, radiation leaks, and death tolls (Unwin, 2019).

With time ticking away and the climate crisis steadily rising in intensity, it seems unthinkable to search outside of the established renewable resources, especially with the success story of renewable energy usage being projected to rise globally in upcoming years. However, the future of energy has been in development for decades, in a technology that could provide limitless energy, skyrocket the planet out of the climate crisis danger zone, and exceed every criterion that current renewables cannot.

Can Cold Fusion Solve our Climate Crisis?

Cold fusion, theoretically, was the holy grail our planet needs to thrive on our growing energy needs while slowing the climate crisis. The original 1989 Fleischmann-Pons press conference promised unlimited, free-flowing energy via the same nuclear fusion processes that produced the Sun's boundless energy – but at room temperature. The implications of this claim are vast. If cold fusion really existed and could be reproduced on a large scale, it would be one of the cleanest forms of energy; no carbon emissions, no radiation or radioactive waste products, and spent materials which can be recycled. Cold fusion would produce endless energy from its cheap and inexhaustible supply of water, which provides the elements necessary for nuclear fusion, like deuterium. It would be constant, steadily generating energy even when the sun is cloaked in clouds and the winds are still. Theoretically, the output of energy characterized by Fleischmann and Pons' observation of "excess heat", when contrasted with the minute space taken up by the input materials, makes the concept of a cold fusion generator extraordinarily powerful, energy efficient, and environmentally friendly (Green Tech Talk, 2018). Theoretically, this should appeal to all... and it did, for a time. There was an insatiable scientific craze around the time of the rushed release of the experiment's results.

The science of how cold fusion theoretically worked created a massive intrigue. To understand, we begin with the concept of nuclear fusion overall. Recall, nuclear energy is currently being produced via fission, where atoms are broken apart and the heat released is harnessed and converted into electricity. Nuclear fusion is exactly the opposite of fission; instead the atoms (typically lightweight hydrogen) are fused together at the nuclei, requiring massive pressure and heat to catalyze the reaction (Green Tech Talk, 2018). As these extreme conditions are found in the plasma of stars, when nuclear fusion is discussed, it can be referred to as "bottling a star on earth" (Kurzgesagt, 2016). It is extraordinarily difficult to force the two positively charged nuclei, which typically repel each other due to like charges, to instead fuse. To replicate conditions of the Sun without the mass of... well, the Sun, scientists have to increase the temperature of fusion beyond the heat of the Sun. This can require temperatures of up to 150 million degrees Celsius, a tremendous amount of energy to force the hydrogen atoms to collide and fuse (Tankler, 2020). In fact, the input of energy is so high, it simply is not feasible to match the output energy to the input, let alone generate a constant surplus of efficient energy for reliable use.

The appeal of cold fusion is the claim that such nuclear fusion can occur at room temperature or below, without the immense energy input, while retaining a spectacular power output. The idea is that when a voltage is applied to the electrodes in an electrolytic solution containing lithium salts and heavy water (water which contains two atoms of deuterium, a heavier hydrogen isotope with two neutrons instead of one), the energy flows from the palladium cathode to platinum anodes. The heavy water molecules in the solution dissociate into anions and cations, where the positively charged deuterium atoms are attracted to the negatively charged palladium strip. Since the palladium atoms are arranged in a lattice, the deuterium atoms squeeze themselves into the interatomic sites, causing the lattice to expand until it is completely saturated. The high deuterium concentration results in synchronized movements throughout the lattice, and eventually the deuterium atoms collide and fuse together under the pressure, releasing a significant amount of heat energy during the process (BlackBoxTechnology, 2010). This theory goes against the commonly accepted laws of physics, and many scientists have debunked Fleischmann and Pons' claims as ludicrous, pathological science.

The secrecy surrounding the methodology of the original experiment means that no replicated model of the fusion currently exists, and the literature largely affirms that the model is purely theoretical. Research on cold fusion is stunted compared to the launch of hot fusion research and other renewable energy development, but nonetheless, it exists, rebranded under one of its less controversial titles: Low Energy Nuclear Reaction (LENR) research.

LENR research is still quietly humming away in the background of the much more widely accepted hot fusion research. Companies are racing to commercialize their research and prove that positive LENR results are reproducible, stirring up skepticism, anger and controversy in a field once thought to be dead. There is, however, an abundance of research and reproducible evidence for a reaction called "muon-catalyzed fusion" which is the closest we've gotten to a legitimate cold fusion reaction.

Muon-catalyzed fusion was actually discovered in 1956, and has been replicably proven as fact, not theory. Muons are heavier versions of electrons, and this mass contributes to a shrunken orbit around the nucleus that allows the muons to bring nuclei approximately 200 times closer together than an electron ever could. With nuclei so close together, fusion becomes so likely that it is inevitable, and can occur at temperatures as low as room temperature. In fact, it has been experimentally proven that these muon-catalyzed fusion reactions are even successful near temperatures of absolute zero (Reich, 2018).

You may now be wondering why muon-catalyzed fusion has not become a mainstream source of energy production. After all, it surpasses all the criteria for renewable energy technology, and unlike cold fusion, it is actually legitimate and proven to work at low temperatures. The problem with this seemingly perfect reaction is the muon that made it all possible in the first place: muons are very energy costly to produce. They have short lifespans, decaying after 2 microseconds, and do not naturally exist in large abundance, so muons have to be produced with a high-energy particle accelerator that requires about 5 gigaelectronvolts (GeV) per muon. Since each muon only produces about 2.7 GeV with these fusions, it consumes more energy than it creates and is thus nonviable for energy production. The only way this could be commercialized and used to fuel human civilization with clean energy is with the rapid advancement of particle accelerators so that less energy is required to produce muons (Reich, 2018).

Currently, as the global energy demand is projected to triple by the end of the century, hot fusion research in the form of takes precedence over replicating cold fusion and muon fusion (ITER, 2019). The experiments are increasingly promising, and with all the current research, a future of cost-effective and abundant energy, sustainability, and pure unpolluted skies may be within reach in the next few decades (ITER, 2019). There may soon be a day where a glass of water can produce as much energy as a whole barrel of oil, but without the environmental repercussions (Kurzgesagt, 2016). There is a global collaboration on the world's largest scale fusion reactor, the International Thermonuclear Experimental Reactor, and years of research are leading to the reactor's first operation in December of 2025 (ITER, 2012). Progressive researchers are even shunning the controversy on cold fusion, with the U.S. Navy scientists recently aiming to "get to the bottom of it" and reveal a reproducible mechanism after years of inconclusive results (Koziol, 2021). A multinational team of scientists in Europe are also recreating the infamous original experiments, investigating cold fusion with modern nanotechnology to develop a new theory and improve upon hydrogen production for fuel cells (Kiger, 2021). NASA has even discovered a method of "confinement fusion" for inducing nuclear fusion in the interatomic spaces of the lattice structure in deuterated metals (Graham & Reckart, 2020).

While cold fusion may only be theoretical, its infamy did help the planet in its own way. It invigorated the never-ending journey to find and commercialize a sustainable source of clean, renewable energy for humanity, dawning a bright future of open minds making pivotal discoveries.

CHAPTER 4

Technologies Involved In Different Fusion Reactions

Wenteng Hou

Technologies involved in cold fusion.

The cold fusion fiasco can be seen as a spectacular case of confirmation bias at best, or fraudulent science at its worst. Most of the world had been heavily focused on the efficiency and improving the technology of nuclear fission reactors, partly motivated by the worst man-made disaster in the history of humankind – Chernobyl, in Soviet Russia, 1986 (Chernobyl Disaster | Causes & Facts, n.d.). The Chernobyl disaster had its impact on the political stage as well as in science, where it probed nuclear physicists and engineers to find safer and less polluting alternatives to nuclear fission. Despite discovery of nuclear fusion since the 1920s and its advancement in military use during the Manhattan Project, controlled fusion reactions that can be harnessed as power have remained elusive until this day (1 November 1952 - Ivy Mike., n.d.). Clearly then, when Fleischmann and Pons announced that they had found a cheaper, more efficient method of achieving controlled nuclear fusion three years after Chernobyl, the physics community exploded. With their simple equipment, it seemed that any individual could find the necessary components in their high school science lab and carry out the exact cold fusion process. The reality, however, was much less optimistic.

This chapter will touch on the technologies required for an actual thermonuclear fusion reactor; the most advanced version is currently in its testing stage in the

International Thermonuclear Experimental Reactor (ITER). The chapter will also go over what Fleischmann and Pons used, their basic electrochemistry principles, and their failures. Finally, technologies associated with the actual "cold fusion", also known as muon-catalyzed fusion, will be discussed.

Requirement of a normal thermonuclear fusion reactor

Ideally, to have a nuclear fusion reaction, you would need the Sun, or any star for that matter. To replicate the exact fusion process occurring within the sun on planet Earth is difficult to say the least. The Manhattan project represents the first time a nuclear fusion reaction had been used by humankind, albeit to a destructive end, through the detonation of Ivy Mike, the thermonuclear hydrogen bomb (1 November 1952 - Ivy Mike., n.d.). To build a nuclear reactor that is capable of producing energy through fusion is another matter entirely.

In its simplest form, nuclear fusion entails the fusion of smaller atoms into a bigger atom. The most known fusion is the proton-proton reaction. This is the fusion of 4 protons into an alpha particle, which is essentially helium but without its electrons, releasing 2 positrons, 2 neutrinos and a burst of energy from the positron-electron annihilation. Overall, this reaction fuses 4 hydrogen nuclei into 1 single helium, a process that is constantly occurring within the Sun (Castelvecchi, 2020). In order to successfully fuse the nuclei, the coulomb barrier must be breached. Because the nuclei are positively charged and like charges repel each other, this electrostatic repulsion energy barrier, named after Coulomb's Law, must be overcome for a fusion reaction to happen. To achieve this within thermonuclear fusion reactors, hydrogen plasma must be used (Fuelling the Fusion Reaction, n.d.). The plasma can then be manipulated in one aspect or another within Lawson's criterion, which describes how temperature, pressure and time affect the nuclear fusion reaction. Plasma is essentially ionized gas state of atoms that can also conduct electricity. Hydrogen in its plasma form is the most common fusion fuel and is produced when hydrogen gas is heated to extremely high temperatures, several million degrees Celsius. This can be achieved through many different heating mechanisms, such as using neutron beam injectors, ion and electron cyclotron resonance heating (Erckmann & Gasparino, 1994). These methods, combined with a changing magnetic field created within the reactor, are able to heat the fuels to high enough temperatures for them to become plasma. This plasma form of hydrogen is then confined in several ways. The most researched and most promising among these are magnetic confinement, through the use of a tokamak that is capable of confining the plasma in the shape of a torus (Tokamak, n.d.). This confinement stops the plasma from cooling, thereby prolonging the reaction so that the plasma can continually fuse with one another. Tokamaks were originally conceived by Soviet nuclear physicists in the 1950s. Since then, they have gained significant traction among the scientific community, with hundreds of tokamaks being built around the world. The largest one is currently under construction at the International

Thermonuclear Experimental Reactor, or ITER, with controlled cooling through a primary water system as well as auxiliary liquid nitrogen and liquid helium cooling systems (Tokamak, n.d.). On the outside of the vacuum seal, ITER uses stainless steel materials for its cryostat, also known as the outer layer, which contains the tokamak within a sealed supercooled vacuum (Cryostat, n.d.). Because ITER uses magnetic confinement to contain the plasma, it is also the place for the world's largest superconducting magnet, built using a niobium-tin and niobium-titanium intermetallic compounds that functions as effective superconductors (Magnets, n.d.). The tokamak themselves are constructed using specific plasma-facing material that must be able to sustain the extremely hot temperatures and the constant bombardment by ions but also able to capture and transmit heat (The Divertor, n.d.). Some examples include boron or silicon carbide and beryllium. There are many other different methods of plasma containment, such as stellarator magnetic confinement or various types of inertial confinements. Despite their differences, they are all attempts to keep the plasma at their desired temperature and prolong the fusion reaction (DOE Explains...Plasma Confinement, n.d.).

The requirement of hydrogen plasma is the primary reason why thermonuclear fusion requires extremely high temperatures. Until today in 2021, no nuclear fusion reactors have been able to achieve break-even. In other words, no reactor have been able to produce enough energy to match the input energy used to run the reactor itself, meaning that the reactor themselves are not self-sustainable yet. In the near future, however, it is hopeful that some of the world's major fusion reactors will begin their trial run and eventually be a fully functioning power source.

Technologies involved in cold fusion.

Because of the sheer difficulty of controlled fusion reaction for the purpose of generating energy mentioned earlier, Fleischmann and Pons shocked the world when these two chemists were thought to have accomplished what the physicists could not for decades. Their version of fusion involved basic electrochemistry and could be accomplished with wide available apparatuses, given they have the right fuel, which is deuterium (Iv et al., 1990).

Deuterium production has been widely used around the world for many decades. In fact, Canada remained the world's largest deuterium producer until 1997, when the nationwide production plants were all shut down (Galley & Bancroft, 1981). The most common way of producing deuterium is by distilling seawater and filtering for heavy water, which is naturally present. Heavy water refers to the compound D_2O, contrasted with the common H_2O. In heavy water, instead of the normal protium, or the hydrogen isotope with no neutron, oxygen is bonded to deuterium, a heavier isotope containing one neutron. The common method for distillation is through the Gridler sulfide process, using two towers, one cold and one hot. This temperature manipulation is to exploit the different equilibrium constants for the reaction between normal water and hydrogen deuterium sulfide into hydrogen deuterium oxide and hydrogen sulfide (Rae, 1978). Huge amounts of deuterium can be produced through this industrial process to supply its many uses in the scientific field.

The basic principle of cold fusion as proposed by Fleischmann and Pons involves the construction of an electrolytic cell. In the most general term, electrolytic cells are electrochemical cells that are supplied by an external power source in order to drive a non-spontaneous reduction/oxidation reaction. This can be used to drive the electrolysis of ionic compounds as well as the process of electroplating, which is similar to what happened within the experiment performed by Fleischmann and Pons. This is contrasted to voltaic cells, which help drive a spontaneous redox reaction. To construct an electrolytic cell, a pair of electrode and electrolytes are required. For the original cold fusion experiment, palladium, in coiled wires, is used at the positive cathode, where the reduction process takes place (Iv et al., 1990). Platinum rod is used for the negative anode, where the oxidation process takes place (Iv et al., 1990). These electrons are then placed in a container filled with lithium hydroxide, LiOD, where deuterium replaces the normal hydrogen in the hydroxide anion. The cathode and anode are then connected to an external source of electricity, such as a battery. The current is passed through this system and the electrolysis process can take place. Oxygen evolution, or the creation of diatomic oxygen O_2, occurs at the anode. Cathode is where the supposed magic happened, as pointed out by Fleischmann in their original announcement. Deuterium was thought to have been liberated from the electrolyte and starts to plate onto the palladium electrode (Iv et al., 1990). These deuterium atoms were said to plate onto the palladium so tightly and densely that their nuclei begin to fuse together. Fleischmann and Pons conjectured that nuclear fusion did happen because they observed an increase in temperature as measured by a calorimeter (Iv et al., 1990). A calorimeter is simply a container that is able to accurately measure the heat released or absorbed within a reaction. There are many different types of calorimeters, ranging from the basic coffee-cup calorimeters to bomb calorimeters that is able to keep a constant volume and withstand large fluctuations in pressure from the reaction. Fleischmann and Pons observed occasional temperature rises from 30 degrees Celsius to 50 degrees Celsius (Iv et al., 1990). They interpreted this as a sign of fusion reaction, which in reality it is not.

The basic equipment used by Fleischmann and Pons was what shocked the scientific community in the first place. The palladium and platinum electrodes, lithium hydroxide with deuterium and calorimeter are all relatively easy to acquire. However, fusion reaction has not been replicated using this simple equipment, and now that this type of cold fusion has been declared as a dead science, despite there are still some people who cling onto the hope that it may work.

Technologies in the 'actual' cold fusion- muon catalyzed fusion.

Despite the Fleischmann – Pons experiment being demonstrated as unachievable, nuclear fusion reaction at room temperature or even lower is very much possible. This has been achieved through muon-catalyzed nuclear fusion. Muons are sub-atomic elementary particles that are similar to electrons, in that they both carry a charge of "-1" and a quantum spin of "1/2". However, muons have a much larger mass, more than 200 times larger than an electron to be more exact (Muon | Sub-atomic Particle, n.d.). It can be influenced by weak nuclear force, or the force that governs nuclear decay. Muons are the result of decay from charged pions, another type of subatomic particle that is made up of exactly one quark and one antiquark. Muons have the potential to be applied in many different areas in science, such as in the industry or medicine, through the use of muon radiographs. However, its extremely limited production does not allow any wide range of applications at the moment, and that includes muon-catalyzed fusion reactors. Normally, muons can be produced via the method where high-energy, high-power proton beams are shone against a stationary target, creating charged pions in the process which then decay into muons (Iiyoshi et al., 2019). Muons are difficult to capture as they are highly unstable, which in turn also makes them difficult to transport.

In a muon-catalyzed fusion reaction, deuterium and tritium are used as fuel and the main fusion reactant, the production process of which have been mentioned above. A beam of muons are sent upon a frozen block of deuterium / tritium mixture, usually kept at a few kelvins above absolute zero. Muons are capable of knocking out the electrons originally contained in the deuterium or tritium atoms (Iiyoshi et al., 2019). Because of its larger mass, the inserted muon between a deuterium and a tritium or between two deuterium atoms is then able to shield the two nuclei from their respective electrostatic repulsion, thereby drawing them closer to each other. Due to this decrease in the distance between the two reactant nuclei, they have a much greater chance of undergoing quantum tunneling and breach through the Coulomb barrier, thus achieving fusion and become an alpha particle (Iiyoshi et al., 2019). The rest of the nuclei are released as energy, which can then be harnessed just as in traditional thermonuclear fusion.

Theoretically, this reaction should be able to proceed infinitely as long as there is deuterium and tritium fuel, but that is not the case in practicality. Muon-catalyzed fusion face a unique problem of alpha-sticking, among others (Jackson, 1957). This means that muons may stick to the alpha particle they helped produce after a while, effectively removing them from the reaction. Unfortunately, this happens much more frequently than expected, which poses a significant challenge to break-even since when the muon catalysts leave, the fusion reaction stops (Jackson, 1957). There is currently no effective method of solving the alpha sticking problem, which is part of the reason why muon-catalyzed reactions are not more widely used.

In conclusion, traditional thermonuclear fusion reactions utilize the creation and confinement of plasma, which is why it is the "hot" fusion as compared to the cold fusion proposed by Fleischmann and Pons. There are many different confinement methods used in experimental fusion reactors around the world, the largest of which is situated at ITER, using a magnetic confinement facility called a tokamak, constructed with giant niobium-tin superconducting magnets. Until this day, fusion reactions have not been able to achieve break-even. These difficulties were part of the reason why Fleischmann and Pons' claimed success in fusion through a simple electrochemistry set up was so shocking yet so unbelievable at the same time. They only required electrodes with somewhat rare metals and deuterium, which is quite abundant and easily extractable. Ultimately, their effort proved to be a spectacular case of confirmation bias in science. There have been other attempts at achieving low temperature fusion, such as through muon catalysis, but the results remain experimental as there are many challenges still to be overcome. Hopefully, in the near future, sustainable nuclear fusion reactors are able to take over the world's energy production.

CHAPTER 5

What does the general public know about cold fusion

Written by Amanda Rande

The development of new technologies within the energy sector are not always met with open arms, there is a large political connection to energy, the political ties and misinformation in the media make it difficult for the average individual to fully understand alternative energy. In the United States the Republicans are known for pushing the use of fossil fuels and rejecting a move into clean alternative energy. This stance closely matches the Conservatives in Canada and directly opposes both the Democrats and Liberals in the United States and Canada respectively. The close affiliation to politics causes a lack of truthful and objective information from reaching the public as news organizations adapt the message and knowledge from scientists to fit their political agenda. The misinformation and vast political ties makes it difficult to discern what the public knows about alternative energy sources and makes the study of media sources an important component for scientific progress. Understanding and analyzing media sources indicates the level of knowledge that the public holds regarding alternative energy sources.

To understand what the public knows about cold fusion technology, a search of popular social media and news sites was conducted. The search was conducted using a general Google search using the keywords cold and fusion. The search resulted in one hundred and fifty-eight million results with many of them being Youtube videos, news articles, and general science information web pages. Two articles were chosen from the first page of results, they were chosen based on their titles as that is how the general public would choose which articles to read. Once the articles were collected a search of social media was undertaken. The popular platforms of Twitter and Reddit were used to gather information pertaining to a larger audience, the popular social media sites have millions of users and provide easy access platforms where communication and knowledge is quickly posted, shared, and saved.

The search of Reddit yielded mixed results with a great number of results relating to a video game and an Adobe computer system. Using the basic Cold Fusion keyword, Reddit supplied three main communities, two of those communities were relating to fusion and nuclear energy while the third was a discussion forum for the Adobe ColdFusion. A further search of the two communities pertaining to nuclear energy were further searched, resulting in videos and posts explaining nuclear fusion.

A search of Twitter using the simple keyword Cold Fusion resulted in no usable results. All results found were relating to the ColdFusion processing system created by Adobe. Based on these results a new search was conducted using the keyword Nuclear Fusion. This search resulted in more accurate data relating to nuclear energy sources; however cold fusion was not the main topic of discussion. This indicates that the public is not getting their knowledge of cold fusion through social media like Twitter, and may not know about or understand cold fusion at all.

The three search engines will be analyzed in tandem to provide an overall view of what the general public understands about cold fusion. This chapter will explore specific Tweets, Reddit posts, and articles found on the front page of Google to indicate the level of understanding held by the public. This is a broad overview and will not delve into the complex differences between political parties or the difference in understanding among classes and races.

Google

Google is the most popular search engine in the world, with over 40,000 searches completed every second and upwards of 3.5 billion completed daily, the search engine amasses incredible amounts of data and has answers to nearly all of life's questions. Being the number one most used search engine, Google provided a solid basis for research into understanding what the public knows about cold fusion. Following a simple Google search using the keywords Cold and Fusion, two articles were selected for analysis. The articles we will analyze were posted/published to Chemical and Engineering News and National Geographic. The first news site is less likely to be accessed by the general public; however, it did appear within the first ten results on the first page of Google and contained a catching title that may draw enough public eyes to indicate the true understanding of the public. The other three websites are rather popular media and educational websites that are frequented by people of all age demographics and education levels.

Article 1: Cold fusion died 25 years ago. but the research lives on

Chemical and Engineering News is an American journal and newspaper published by the American Chemical Society, and upholds a scientific viewpoint and stance as it only publishes technical news and field analyses. The journal itself is not readily accessible or likely to be read by the general public, but the location of this article in the Google generated results and the catchy title indicate that it has been viewed by a larger audience.

The article uses a lot of technical jargon without explaining what it means, it is clearly not aimed at the general public. However, it does provide an overview of the history and how cold fusion came to be. Ritter (2016) explains that Randell L. Mills developed a theory in 1991 stating that hydrogen's electron can be lowered to a more stable state that produces copious amounts of heat energy. "Fleischman and Pons said this process could not be caused by any known chemical reaction, and the nuclear reaction term 'cold fusion' was attached to it" (Ritter, 2016). Ritter says that this theory was developed following a movement in the 1980s to create a small scale nuclear fusion device that could sit on benchtops and supply nearly unlimited power. He continues by outlining the failures of cold fusion, as the production of heat energy could not be reliably reproduced in laboratory settings using Mill's theory causing the research to be " summarily condemned, and cold fusion became a synonym for junk science" (Ritter, 2016).

Following Ritter's introduction to the development and failure of cold fusion technologies he discusses current science and research models, introducing a BLP, a company currently testing the SunCell. The SunCell is a device that uses a similar theory to Mills' to split hydrogen and create energy. So far the research has not been proven and is not ready for mass production. Many scientists researching the topic are no longer active in the scientific community and have since retired from their careers. The retirement of the lead scientists and a lack of funding has stalled the development of cold fusion or cold fusion like technology. Ritter (2016) goes on to explain that the bad reputation cold fusion garnered in the 90s has created social stigma that has discouraged companies from investing in further research and production.

Although not explicitly written for the general public, the scientific jargon and breakdown of key events in the development and abandonment of cold fusion provides key insights for the public to understand how and why cold fusion initially occurred and quickly failed.

Article 2: Cold fusion remains elusive -but these scientists may revive the quest

National Geographic is a popular magazine that focuses on natural and social sciences, they are owned by Disney and produce dozens of shows, movies, documentaries, and articles each year. Their focus on science and long history of being a reliable source makes them a good resource for individuals who want to learn more about natural science topics.

Greshko's (2019) article focuses on the current resurgence of cold fusion technology that is being funded by Google. He begins by explaining how cold fusion uses the same nuclear reaction that occurs in the sun, but it occurs at room temperature making it a much safer and viable alternative to regular nuclear energy. He briefly touches on the failures of cold fusion, stating that it was "swiftly labeled a lost cause" and research stopped almost as abruptly as it had started. Greshko (2019) touches on the science behind nuclear energy, explaining it in terms that are digestible by anyone with a high school level science education. He goes on to explain why fusion of all kinds is difficult, saying "realizing fusion power is possible at high densities and temperatures, if the nuclei are confined for sufficiently long time" and noting that the equipment, time, and manpower required to confine the nuclei for long enough is incredibly costly.

The article then jumps to 2015, when Google began funding a group of researchers to revisit the idea of cold fusion. Greshko (2019) touches on the rigorous peer-review process this new group of scientists underwent to ensure that any data they compile is accurate and repeatable. He speaks to the three research projects they have completed and how the new data garnered from these studies is wildly different from those studies published by the likes of Mills, Pons, and Fleischamn in the late 80s and 90s. He wraps up the article by expressing hope for future experiments and research studies to replicate cold fusion and fully determine if it is real or not.

This National Geographic piece provides detailed yet digestible scientific explanation, discussions and quotes from scientists, and hope for the future. It is an article that the public may rely on for information regarding cold fusion and provides them with new knowledge while explaining historical studies.

Twitter

Twitter is a social media platform that connects millions of individuals through 250 character posts/tweets. People speak their minds, communicate with friends and family, find new hobbies, and conduct political campaigns all through this simple platform. In 2020, 186 million people had active Twitter accounts. This large number of active users makes Twitter a prime location to learn information and post questions and articles about the newest scientific discoveries. Completing a simple search using the keywords Cold Fusion resulted in thousands of tweets regarding a program by Adobe. THis information does not relate or connect to nuclear fusion at all, therefore, a second search was conducted using the keywords Nuclear fusion, this search yielded thousands of results relating to nuclear power plants and nuclear destruction. A user account titled Nuclear Fusion Project had the greatest number of tweets under this search and is the basis for this section's analysis. The Nuclear Fusion Project's bio reads "A new pipeline providing networking, mentorship, and publication assistance to aspiring national security professionals", thus indicating that the account will not be directly associated with cold fusion.

The lack of information and tweets regarding cold fusion on Twitter may indicate a lack of nuclear directed scientific minds on the platform, or may be indicative of a lack of understanding or interest in the public who use this platform. Being one of the leading social media platforms it is important to consider and understand the role Twitter plays within the public consciousness and how information or lack thereof is communicated among its vast user base.

Reddit

Reddit is a social media platform that functions through subpages with each sub-page containing a different topic. Within each subreddit there may be thousands of posts or conversations among users, with 222 million users last year, Reddit makes up one of the most popular social media platforms for self proclaimed nerds and geeks. Some of the most popular subreddits relate to philosophy, coding, mo-lecular sciences, and many other scientific disciplines. This makes Reddit unique in that it is the most likely to have genuine conversations and discussions about cold fusion by both normal everyday individuals and highly educated scientists.

The search began by inputting r/coldfusion into Reddit's search engine (r/ is reddits way to determine and differentiate subreddits and subpages). The general search yielded thousands of results, ranging from subreddits dedicated to cold fusion, low-energy nuclear reactions (LENR), fusion, and posts with hashtags or content relating to cold fusion.Due to the broad variety of topics presented on Reddit, there were thousands of responses, posts, and groups relating to the ColdFusion software created by Adobe, all results pertaining to that subject have been left out; however, it is important to note that the names do cause some confusion within the comment section of some posts. There are many individuals who will speak of the nuclear reaction on posts relating to the software and vice versa. For the purpose of this chapter, we will focus on two posts relating to cold fusion and the subreddit r/LENR.

The first post we will analyze is a simple questionnaire, the poster (u/unspecial-noob) wrote "Wyr end world hunger, have world peace, or solve cold fusion", in this instance Wyr is an abbreviation for would you rather and the three options were selectable by viewers of the subreddit. The post received twelve comments and 498 votes. Of those 498, 168 people voted to solve cold fusion. User Beled-agnir commented on the post saying "Cold fusion would go a long way towards solving the issue of hunger/resources...", thus indicating an understanding as to how the application of cold fusion would benefit the world as a whole. Anoth-er commenter, user Billy999mays asked "whats cold fusion", thus showing that within a single subreddit there is a vast difference in understanding of the topic. Their comment was answered by the poster of the initial ``what would you rather' explanation that cold fusion is saying "a method of nuclear power. It is largely considered the most perfect method of power generation possible under known physics". This response does reflect a deeper knowledge and understanding of cold fusion, and does align with the opinions presented by several nuclear scientists who are pushing to continue the research and development into cold fusion.

The second post we will analyze was written by user YelirRuessvel and simply asks what cold fusion is. This post garnered a lot less attention than the previous, but did receive eight comments explaining what cold fusion is. The first two comments on the post took two different routes to answering the question. The first commenter gave a simple and brief answer where the defined nuclear fusion then cold fusion. The second commenter went into great detail regarding how nuclear fusion works and why cold fusion failed. These two posts are important for observing the overall understanding of the public; however, there was a string of comments and responses to user r4d4r_3n5's comment "Something that doesn't work…" that provide a broader view of what the people of REddit know about cold fusion. The first two commenters spoke about the difficulties of bringing two atoms close enough together to fuse them in a safe and effective manner. User DavidRFZ points out that current scientists are able to generate enough power for scientific analysis or to detonate a hydrogen bomb, but have not yet found a way to harness the energy in the middle. This comment shows a great understanding of both hydrogen bombs and nuclear fusion at many different levels. User r4d4r_3n5 was the final commenter, they explained that Fleishman and Pons announced a technology and a theory that other scientists could not reliably reproduce, they mention the necessity of the scientific method while crediting the ground breaking finding by Fleishman and Pons. This response shows a great knowledge of both the history, creation, development, and failures of cold fusion.

The subreddit r/LENR provides a catch all location for individuals interested in learning more about or discussing lower energy nuclear reactions. FOllowing the failure of Fleishman and Pons, the scientific community rebranded cold fusion as low energy nuclear reactions. This name both explains what is occuring during cold fusion while avoiding the negative associations that the scientific community and the media have with cold fusion. The newest post on r/LENR is a youtube video that discusses the relationship, observations, and explanations of LENR. This educational video indicates a greater knowledge base and understanding of nuclear physics than all other areas of the internet we explored in this chapter.

Conclusion

The understanding of the general public regarding cold fusion is broad and far reaching with some individuals having a great deal of knowledge that they are able to pass onto others while some have never heard of cold fusion and confuse it for the Adobe software program. This chapter analyzed three main search engines/platforms to provide an overview of the general public's understanding and thoughts on cold fusion. There are many accessible resources written in both scientific jargon and normal everyday language that can increase people's understanding of cold fusion. Based on the results from the Google search, it can be theorized that a large number of people have a base understanding of cold fusion or have at least heard the term in relation to alternative energy. The lack of results found on Twitter may indicate a lack of knowledge and interest within the social media platform's user base or may simply represent a platform whose focus is elsewhere. The search of Reddit yielded the greatest results and the greatest discussions between its users. This indicates that there is a general to informed understanding of cold fusion within the public. This chapter is limited to the three search engines/platforms analyzed and does not cover the public's knowledge as a whole; however it does provide insights into what the public thinks of cold fusion and where they learn and discuss the topic the most.

CHAPTER 6

The Future of Cold Fusion

By Alexander Martin

Cold fusion was a concept that was first introduced in 1989 by Professor Stanley Pons and Martin Fleischmann. Their initial report, which claimed that they achieved fusion at room temperature, enthralled scientists. In a short synopsis, the two hypothesized that compression and deuterium combined with palladium metal while using electrolysis resulted in nuclear fusion Their claims fascinated many because the process could create a new form of renewable energy (Schwartz, 2020). The potential of cold fusion had many trying to replicate the experiment to validate and verify the claims made by Fleischmann and Pons. Few were able to find any substantial evidence of cold fusion, resulting in the retraction of their claims. The team relocated their laboratory to France to continue their research (Pilkington, 2003). However, the possibility of cold fusion continued to interest many Since the initial announcement, efforts have been made to advance the field and expand on research. The concept of fusion will be explored in the following chapter along with Google's current research into cold fusion. Various other companies are continuing to pursue the topic of cold fusion, including the U.S. government, NASA, and Toyota. Findings in the field will be discussed and the stigma surrounding cold fusion. Overall, cold fusion is a prominent topic explored by various companies, governments, etc., suggesting its potential in the future.

What is fusion?

Nuclear Fusion is a subatomic reaction that occurs when the strong nuclear force between two atoms becomes so great that the two atoms merge together and form an atom of a heavier element (Conn, 2019). Nuclear fission is the process where heavier elements are split into lighter elements either by particles decaying naturally over time, or through the bombardment of subatomic particles onto an atom, usually of a heavier element (Conn, 2019). Through these processes, the particles produce a significant amount of energy. Notably, fusion is what powers all stars in the universe, our sun included (Conn, 2019). The atomic bombs dropped on Hiroshima and Nagasaki in 1945 utilized exclusively fission. Also, consider that "the only nuclear energy [that] can be [controlled] is fission, which is what nuclear power plants use" (Scientific American, 1999). Both these processes require high temperatures up to 100 million Kelvin, which is approximately six times hotter than the sun's core (Source 2). On the other hand, "cold fusion" is considered 'cold' because the reaction can occur at room temperature.

Google's Experiments with Cold Fusion

The technology company Google, invested in the idea of cold fusion by funding a group of 30 researchers to study the concept since 2015. Google disrupted the research across multiple laboratories, investing 10 million dollars into the topic. The researchers were divided into teams across various academic institutions such as MIT, UBC, and the University of Maryland (Gibney, 2019). Notable scientists participated in the research including physicist Thomas Schenkel from the Lawrence Berkeley National Laboratory. Others involved included 30 graduate students, and multiple postdoctoral researchers. The team's goal was to see if any group was able to locate any characteristics associated with cold fusion. The "... experiment aimed to address a key claim within the cold fusion community: If at least seven for every eight palladium atoms – the device gives off excess heat. But as the researchers soon realized, packing palladium full of deuterium is extremely difficult, and so is measuring it" (Greshko, 2019). The conditions were hard to recreate since the original instance of cold fusion is considered a phenomenon by most scientists. Considering this, not a lot of information is available concerning cold fusion. Thus, researchers needed to experiment and determine the best way to potentially create cold fusion. Having a large group of researchers working together, "If credible evidence of an anomaly were found, the apparatus would be developed into a reference experiment that could be vetted by the rest of the peer group and eventually the broader scientific community" (Greshko, 2019). The teams created various experiments that could recreate the original conditions found by Fleischmann and Pons in 1989. The teams attempted to create thermal energy by electrolysis. Here, heavy water and a palladium cathode (a catalytic converter) were used together to see if the energy created from their pairing exceeded the original inputted amount. Furthermore,

> "The researchers pursued the three experimental strands that they deemed sufficiently credible. In one, they tried to load palladium with amounts of deuterium hypothesized to be necessary to trigger fusion. But at high concentrations the team was unable to create stable samples" (Gibney, 2019).

This was the first situational experiment and Google conducted two more in addition. In the second situation, a team utilized work done in the 1990s, which claimed to have generated amounts of tritium during a cold fusion experiment. This supposedly occurs when the palladium is bombarded with pulses of hot deuterium ions (Gibney, 2019). After recreating the conditions, Google was unable to produce the same outcomes. Two of Google's experimental models have failed but they still completed a final third version. The final experiment type "involved heating up metallic powders in a hydrogen-rich-environment. Some current proponents of cold fusion claim that the process produces excess and unexplained heat, which they theorize is the result of fusing elements" (Pilkington, 2003). However, with nearly 420 tests, the Google-funded team found no such heat excess (Gibney, 2019). After the fact, they released a statement that "So far, [Google has] found no evidence of anomalous effects claimed by proponents of cold fusion" (Gibney, 2019). Considering this, Google's efforts were unable to create the conditions necessary for cold fusion to occur. Further research is being conducted by the teams as they believe there is greater potential of exploring instances of cold fusion in the future. For instance, some of their technology could have potentially not picked up on anything. The researchers believe that the second situational experiment involving tritium "could be too small to measure with current equipment…" (Gibney, 2019). As technology continues to advance, potentially researchers can come back to the subject and explore the potential for cold fusion. In the meantime, Google is going to continue working with TAE Technologies, a company that focuses on developing clean energy (Greshko, 2019). Overall, Google is highly invested in the idea of cold fusion. As a company, they have provided a group of researchers funding to see if instances of cold fusion can be discovered. Although the group was unable to find such occurrences so far, Google is continuing to fund them going forward. Therefore, Google is working on understanding the intricacies of cold fusion.

U.S. Navy Researchers Explore Cold Fusion

Currently, the Naval Surface Warfare Center, Indian Head Division is working together with a group of the U.S. Navy and the National Institute of Standard and Technology (NIST) (Koziol, 2021). The group is working together to reveal the potential of cold fusion as a reusable efficient energy source (Koziol, 2021). Together, there are five laboratories that are working on the subject to reveal if cold fusion can be achieved and if it can be used as an energy source. The team is working to publish their results but there is no clear date that the research will be done.

NASA and Nuclear Fusion

Although not achieving fusion at room temperature like cold fusion, NASA has found success with related experiments. Similar to the cold fusion involving deuterium, the NASA experiment involves something called "lattice confinement."

This is where "fusion takes place in the narrow channels between atoms," and where energy is created as a result (Delbert, 2021). More specifically, this occurs when deuterium gets caught in a solid metal which creates an effect that is neither supercooled nor superheated, "...but where atoms reach fusion-level energy" (Delbert, 2021). NASA is exploring this as a potential energy source for space missions. Describing the lattice confinement process, NASA notes:

> *"In the new method, conditions sufficient for fusion are created in the confines of the metal lattice that is held at ambient temperature. While the metal lattice, loaded with deuterium fuel, may initially appear to be at room temperature, the new method creates an energetic environment inside the lattice where individual atoms achieve equivalent fusion-level kinetic energies"* (Delbert, 2021).

This process is achieved with the help of ITER, an international nuclear fusion research project aimed at replicating the energy created by the sun. ITER has helped to produce a device called the Tokamak which is located in southern France (Delbert, 2021). The device itself started development in 2013 and the first run to test nuclear fusion is projected to take place in 2025. Notably, the cost of this device will exceed $60 billion Euros (Delbert, 2021). Electromagnetic confinement is the most studied category in the world when it comes to nuclear fusion. However, another version that is not always considered is called inertial confinement fusion, which may be the future (Windridge, 2020). Lattice confinement is a form of nuclear fusion in which deuterium is interspersed and confined between atoms of a metallic solid. NASA is developing this technology to be applicable to their rockets (Delbert, 2021). Through the process of lattice confinement, it is believed plasma is a form of possible rocket fuel. Again, nuclear fusion has the potential to produce energy. A part of this process, lattice confinement utilizes deuterium which is applied between the atoms of a solid metal at room temperature. This in turn is a way of producing energy for NASA's rockets going forward (Delbert, 2021). The future of nuclear fusion as a form of energy is continuing to develop. Cold fusion touched on this idea in the 1980-1990s with the Fleischmann and Pons experiment. Explorations of fusion related concepts are continued to be explored with NASA's work within the subject.

Toyota's Work in the Field

The automotive company Toyota has been working to develop the area of research that focuses on cold fusion. Toyota established a new organization called IMRA which has "...two laboratories, one near Sapporo and the other near Nice in the south of France..." ("Scientific American," 1999). The company IMRA has funding from Toyota and has created a calorimeter, a device that measures the physical heat produced by chemical reactions. Using the calorimeter, the company conducted various experiments that focused on producing a cold fusion at room temperature. However, no success was found. It was noted that, "Twenty-six experiments were tried employing various systems and tricks that had been suggested to cause excess heat, but no excess heat was observed ("Scientific American," 1999). Further, the upper limits were very low, +/- 0.23 watts out' and the hundreds of percent increases claimed back in 1989. Considering the experiment conducted by IMRA, the company was unsuccessful in producing cold fusion ("Scientific American," 1999). The company continues to fund the two labs which are continuing to explore the potential for cold fusion as a possible renewable energy.

Various Findings

After working in the United States, both Pons and Fleischmann went to France to work with a group under G. Lonchampt with the support of the French Atomic Energy Commission. During their experiments, some reports found that "excess powers of 100 watt (150 percent of input power) sustained over a 30-day run" ("Scientific American," 1999). Notably, "Most cold fusion research today is done in Japan. The New Energy and Industrial Technology Development Organization, a government organization, sponsors the New Hydrogen energy laboratory in Sapporo. IMRA, a foundation of the Toyota family, sponsors another well-equipped lab in Sapporo, as well as the Pons and Fleischmann's facility in France. Several Japanese universities and industries also do cold fusion research. ("Scientific American," 1999). There could be possible findings in the future that help to better understand the phenomenon. Japan continues to do effective research on the topic of cold fusion. Notably, the University of Hokkaido has been analyzing "...the components of a Pd-heavy water cell before and after an extended run at high temperature. They reported low concentrations on a range of heavy elements, including calcium, titanium, chromium, manganese, iron, copper, and zinc" ("Scientific American," 1999). It is uncertain what is taking place in these reactions to eliminate metals found priorly in the water. Potentially there is a process that is taking place that is not detectable by modern practices ("Scientific American," 1999). All considered, work on nuclear fusion continues and the investigation of the possibility of cold fusion is still underway, future advances in technology may change people's understanding of the concept and whether or not it is possible.

The Issue with the Future of Cold Fusion

The issue with the future of cold fusion is the fact that a lot of scientists believe it is not possible. Without having the original hypothesis properly tested, people in the field believe that it is a pipedream that cannot be executed. After the Pons and Fleischmann experiment, various scientists tried to replicate cold fusion but little to none have been successful. As a result, "...most scientists have not followed the field since the disenchantment of 1989 and 1990 ("Scientific American," 1999). They typically still dismiss cold fusion as experimental error, but most of them are unaware of newly reported results. Even so, given the extraordinary nature of the claimed cold fusion results, it will take extraordinarily high quality, conclusive data to convince most scientists, unless a compelling theoretical explanation is found first" ("Scientific American," 1999). This is part of the problem with the future of cold fusion. Many scientists do not take the idea seriously, which as a result limits research. With a prejudice surrounding the topic, people are discouraged from participating in cold fusion because of this. Again, Japan is a place where cold fusion is taken seriously. Notably, in 1996, Japan hosted "The Sixth International Cold Fusion Conference" which was sponsored by the Japanese organization MITI ("Scientific American," 1999). Since then, MITI has provided Japan with $30 million in funding fusion research in the country. As a result, other Japanese companies proceeded to match the funding provided by MITI to excel the field of research ("Scientific American," 1999). This included research conducted by Jirohta Kasagi of Tohoku University in Japan. His experiments attempted to recreate Fleischmann and Pons hypothesis. He conducted the following test:

> "[First] deuterium ions of a variety of low energies were fired into metals that had been saturated with deuterium; the measured rates of fusion were then compared with expectations. The rates decreased steeply at low energies because of the Coulomb barrier (electrical repulsion), and no unexpected enhancement was observed of the kind that would be needed to justify Fleischmann and Pons's claims" ("Scientific American," 1999).

With nearly $10 million dollars of equipment at their disposal, the Japanese researchers and scientists could not produce cold fusion ("Scientific American," 1999). Overall, Japan has the potential to reveal the intricacies of cold fusion. Overall, with excessive funding, Japan is working to progress research and better understand the idea of cold fusion as a whole.

Conclusion

Cold fusion is a topic that has a lot of energy being put into its development. Companies such as Google are investing a lot of resources into finding results concerning the phenomenon. So far, there have been no significant results from the lab since 2015. However, the company is confident that it is potentially not picking up on processes due to a lack of technology. Future technology can help develop scientists' understanding of the concept. Japan has invested in the sector, conducting prevalent research. As a whole, the country is progressing scientists' understanding of the concept. To this day, nobody has been able to replicate what Fleischmann and Pons claimed to find. However, serious research is being conducted regarding cold fusion. Scientists' understanding of the concept will only continue to advance.

REFERENCES

Chapter 1

Bohr, N. (1923). The Structure of the Atom 1. Nature, 112(2801), 29–44. https://doi.org/10.1038/112029a0

Cold fusion: A case study for scientific behavior. (2012). https://undsci. berkeley.edu/article/cold_fusion_01

Graham, T. (1869). On the relation of hydrogen to palladium. Proceedings of the Royal Society of London, 17, 212–220. https://doi.org/10.1098/rspl.1868.0030

Greshko, M. (2019, May 29). Cold fusion remains elusive—But these scientists may revive the quest. Science. https://www.nationalgeographic.com/science/article/cold-fusion-remains-elusive-these-scientists-may-revive-quest

Homlid, L. (2019). Existing Source for Muon-Catalyzed Nuclear Fusion Can Give Megawatt Thermal Fusion Generator. Fusion Science and Technology, 75(3), 208–217. https://doi.org/10.1080/15361055.2018.1546090

Krivit, S. B. (2009). ENERGY | Cold Fusion: History. In J. Garche (Ed.), Encyclopedia of Electrochemical Power Sources (pp. 271–276). Elsevier. https://doi.org/10.1016/B978-044452745-5.00945-X

Lewenstein, B. V. (1992). Cold Fusion and Hot History. Osiris, 7, 135–163.

Murdoch, H. (1989, May 11). Cold Fusion? It Was First Tried in 1927. AP NEWS. https://apnews.com/article/7c7bd232f39e9c8c34f94305c435bcf7

Chapter 2

Basic fusion physics | IAEA. (2016, October 12). Iaea.org. https://www.iaea.org/topics/energy/fusion/background

Benefits of Fusion - What is fusion energy and why do we need it? (2021, June 16). General Fusion. https://generalfusion.com/what-are-the-benefits-of-fusion-energy/

Delbert, C. (2021, January 25). Jeff Bezos Is Backing an Ancient Kind of Nuclear Fusion.

Popular Mechanics; Popular Mechanics. https://www.popularmechanics.com/science/energy/a35276502/magnetized-target-fusion-technology-jeff-bezos/

EUROfusion. (2017). History of Fusion- EUROfusion. Euro-Fusion.org. https://www.euro-fusion.org/fusion/history-of-fusion/

EUROfusion. (2020). JET- EUROfusion. Euro-Fusion.org. https://www.euro-fusion.org/devices/jet/

Fusion reactor - Mirror confinement | Britannica. (2021). In Encyclopædia Britannica. https://www.britannica.com/technology/fusion-reactor/Mirror-confinement

McGrath, M. (2021, June 17). Nuclear energy: Fusion plant backed by Jeff Bezos to be built in UK. BBC News; BBC News. https://www.bbc.com/news/science-environment-57512229

Morse, E. (2018). Nuclear Fusion (1st ed. 2018.). Springer International Publishing.https://doi.org/10.1007/978-3-319-98171-0

Nuclear fusion promises a clean, green energy future — but there's a catch - ABC News. (2020, April 7). ABC News. https://www.abc.net.au/news/2020-04-08/future-tense-nuclear-fusion-sustainable-power-promise/12114948

Nuclear Fusion | Development, Processes, Equations, & Facts | Britannica. (2021). In Encyclopædia Britannica. https://www.britannica.com/science/nuclear-fusion

Nuclear Fusion: WNA - World Nuclear Association. (2021). World-Nuclear.org. https://www.world-nuclear.org/information-library/current-and-future-generation/nuclear-Fusion-power.aspx

Nuclear Fission | Boundless Chemistry. (2021). Lumenlearning.com. https://courses.lumenlearning.com/boundless-chemistry/chapter/nuclear-fission/

Nuclear Bombardment Reactions - Assignment Point. (2021). Assignmentpoint.com. https://www.assignmentpoint.com/science/chemistry/nuclear-bombardment-reactions.html

US EPA,OAR. (2015, April 15). Radionuclide Basics: Tritium | US EPA. US EPA. https://www.epa.gov/radiation/radionuclide-basics-tritium

Chapter 3

BlackBoxTechnology. (2010, August 12). Cold Fusion: How it works. Www.youtube.com. https://www.youtube.com/watch?v=f6d2q-YxVvk

Demirtas, O. (2013, December). Evaluating the Best Renewable Energy Technology For Sustainable Energy Planning. ResearchGate. https://www.researchgate.net/publication/285773953_Evaluating_the_Best_Renewable_Energy_Technology_For_Sustainable_Energy_Planning

Government of Canada. (2016). Renewable energy facts | Natural Resources Canada. Nrcan.gc.ca. https://www.nrcan.gc.ca/science-data/data-analysis/energy-data-analysis/energy-facts/renewable-energy-facts/20069

Government of Canada. (2017). Energy and Greenhouse Gas Emissions (GHGs) | Natural Resources Canada. Nrcan.gc.ca. https://www.nrcan.gc.ca/science-data/data-analysis/energy-data-analysis/energy-facts/energy-and-greenhouse-gas-emissions-ghgs/20063

Graham, S., & Reckart, T. (2020, April). Lattice Confinement Fusion. Glenn Research Center | NASA. https://www1.grc.nasa.gov/space/science/lattice-confinement-fusion/

Green Tech Talk. (2018, August 3). What is Cold Fusion? Green Tech Talk. https://www.greentechtalk.com/what-is-cold-fusion/

Green Tech Talk. (2019, May 15). How Does Non Renewable Energy Affect The Environment? Green Tech Talk. https://www.greentechtalk.com/how-does-non-renewable-energy-affect-the-environment/

ITER. (2012). What will ITER do ? ITER. https://www.iter.org/sci/Goals

ITER. (2019). Advantages of fusion. ITER. https://www.iter.org/sci/Fusion

Kiger, P. (2021, January 22). Is the Dream of Cold Fusion Still a Possibility? HowStuffWorks. https://science.howstuffworks.com/environmental/energy/cold-fusion.htm

Koziol, M. (2021, March 22). Full Page Reload. IEEE Spectrum: Technology, Engineering, and Science News. https://spectrum.ieee.org/tech-talk/energy/nuclear/cold-fusion-or-low-energy-nuclear-reactions-us-navy-researchers-reopen-case

Kurzgesagt. (2016). Fusion Power Explained – Future or Failure. In YouTube. https://www.youtube.com/watch?v=mZsaaturR6E

Office of Nuclear Energy. (2021, March 31). 3 reasons why nuclear is clean and sustainable. Energy.gov. https://www.energy.gov/ne/articles/3-reasons-why-nuclear-clean-and-sustainable

Reich, H. (2018). Legitimate Cold Fusion Exists | Muon-Catalyzed Fusion [YouTube Video]. In YouTube. https://www.youtube.com/watch?v=aDfB-3gnxRhc

Tankler, A. (2020, November 26). what is cold fusion? European Investment Bank. https://www.eib.org/en/stories/renewable-energy-source-cold-fusion

Union of Concerned Scientists. (2013, March 5). Environmental Impacts of Renewable Energy Technologies. Union of Concerned Scientists. https://www.ucsusa.org/resources/environmental-impacts-renewable-energy-technologies

Unwin, J. (2019, May 28). Nuclear power pros and cons: What's the impact of the energy source? Power Technology | Energy News and Market Analysis. https://www.power-technology.com/features/nuclear-power-pros-cons/

What are some of the signs of climate change? (n.d.). Www.usgs.gov. https://www.usgs.gov/faqs/what-are-some-signs-climate-change-1?qt-news_science_products=0#qt-news_science_products

Chapter 4

1 November 1952—Ivy Mike. (n.d.). CTBTO Preparatory Commission. Retrieved June 25, 2021, from https://www.ctbto.org/specials/testing-times/1-november-1952-ivy-mike

Castelvecchi, D. (2020). Neutrinos reveal final secret of Sun's nuclear fusion. Nature, 583(7814), 20–21. https://doi.org/10.1038/d41586-020-01908-2

Chernobyl disaster | Causes & Facts. (n.d.). Encyclopedia Britannica. Retrieved June 25, 2021, from https://www.britannica.com/event/Chernobyl-disaster

Cryostat. (n.d.). ITER. Retrieved June 25, 2021, from http://www.iter.org/mach/cryostat

DOE Explains...Plasma Confinement. (n.d.). Energy.Gov. Retrieved June 25, 2021, from https://www.energy.gov/science/doe-explainsplasma-confinement

Erckmann, V., & Gasparino, U. (1994). Electron cyclotron resonance heating and current drive in toroidal fusion plasmas. Plasma Physics and Controlled Fusion, 36(12), 1869–1962. https://doi.org/10.1088/0741-3335/36/12/001

Fuelling the Fusion Reaction. (n.d.). ITER. Retrieved June 25, 2021, from http://www.iter.org/sci/fusionfuels

Galley, M. R., & Bancroft, A. R. (1981). Production d'eau lourde canadienne—De 1970 a 1980. Atomic Energy of Canada Limited, 112.

Iiyoshi, A., Kino, Y., Sato, M., Tanahashi, Y., Yamamoto, N., Nakatani, S., Tendler, M., & Motojima, O. (2019). Muon catalyzed fusion, present and future. AIP Conference Proceedings, 8.

Iv, J. S. B., Powell, G. L., & Hutchinson, D. P. (1990). Electrochemical factors in cold fusion experiments. Journal of Fusion Energy, 6.

Jackson, J. D. (1957). Catalysis of Nuclear Reactions between Hydrogen Isotopes by ${\ensuremath{\mu}}^{\ensuremath{-}}$ Mesons. Physical Review, 106(2), 330–339. https://doi.org/10.1103/PhysRev.106.330

Magnets. (n.d.). ITER. Retrieved June 25, 2021, from http://www.iter.org/mach/magnets

Muon | subatomic particle. (n.d.). Encyclopedia Britannica. Retrieved

June 25, 2021, from https://www.britannica.com/science/muon

Rae, H. K. (1978). Separation of Hydrogen Isotopes. American Chemical Society, 26.

The divertor. (n.d.). ITER. Retrieved June 25, 2021, from http://www.iter.org/mach/divertor

Tokamak. (n.d.). ITER. Retrieved June 25, 2021, from http://www.iter.org/mach/tokamak

Chapter 5

u/Unspecialnoob. (n.d.). Wyr end world hunger, have world peace, or solve cold fusion? [Content]. Reddit. https://www.reddit.com/r/WouldYouRather/comments/n4ak7p/wyr_end_world_hunger_have_world_peace_or_solve/

u/Billy999mays. (n.d.). Whats cold fusion. [Content]. Reddit https://www.reddit.com/r/WouldYouRather/comments/n4ak7p/wyr_end_world_hunger_have_world_peace_or_solve/

u/YelirRuessvel. (n.d.). Eli5 What's cold fusion? [Content]. Reddit. https://www.reddit.com/r/explainlikeimfive/comments/myk4td/eli5_whats_cold_fusion/gvvdtmm/

u/DavidRFZ. (n.d.). The purpose of a power plant is to generate heat which boils water to create steam which is sent into a turbine which creates electricity. [Description]. Reddit. https://www.reddit.com/r/explainlikeimfive/comments/myk4td/eli5_whats_cold_fusion/gvvdtmm/

u/r4dr_3n5. (n.d.). Just look up Pons & Fleischmann. [Description]. Reddit. https://www.reddit.com/r/explainlikeimfive/comments/myk4td/eli5_whats_cold_fusion/gvvdtmm/

Greshko, M. (2019). Cold fusion remains elusive - but these scientists may revive the quest. National Geographic. https://www.nationalgeographic.com/science/article/cold-fusion-remains-elusive-these-scientists-may-revive-quest

Ritter, S.K. (2016). Cold fusion died 25 years ago, but the research lives on: Scientists continue to study unusual heat-generating effects, some hoping for vindication, others for an eventual payday. Chemical and Engineering News, 94(44).

Chapter 6

Delbert, C. (2021, March 10). NASA Found Another Way Into Nuclear Fusion. Popular Mechanics. https://www.popularmechanics.com/science/energy/a34096117/nasa-nuclear-lattice-confiment-fusion/.

Dowling , S. (2017, August 16). The monster atomic bomb that was too big to use. BBC News https://www.bbc.com/future/article/20170816-the-monster-atomic-bomb-that-was-too-big-to-use.

Greshko, M. (2021, May 3). Cold fusion remains elusive-but these scientists may revive the quest. National Geographic. https://www.national-geographic.com/science/article/cold-fusion-remains-elusive-these-scientists-may-revive-quest.

Koziol , M. (2021, March 22). Whether Cold Fusion or Low-Energy Nuclear Reactions, U.S. Navy Researchers Reopen Case. IEEE Spectrum. https://spectrum.ieee.org/tech-talk/energy/nuclear/cold-fusion-or-low-energy-nuclear-reactions-us-navy-researchers-reopen-case.

Pilkington , M. (2003, April 10). Cold fusion. The Guardian https://www.theguardian.com/science/2003/apr/10/farout.

Scientific American. (1999, October 21). What is the current scientific thinking on cold fusion? Is there any possible validity to this phenomenon? https://www.scientificamerican.com/article/what-is-the-current-scien/.

Schwartz , O. (2020, December 28). Is nuclear fusion the answer to the climate crisis? The Guardian. https://www.theguardian.com/environment/2020/dec/28/nuclear-fusion-power-climate-crisis.

Windridge, M. (2020, October 9). The New Space Race Is Fusion Energy. Forbes. https://www.forbes.com/sites/melaniewindridge/2020/10/07/the-new-space-race-is-fusion-energy

www.ingramcontent.com/pod-product-compliance
Lightning Source LLC
Chambersburg PA
CBHW031815190326
41518CB00006B/342